Voyage of the 3I-ATLAS

A Messenger from the Stars

By

Brian Keith Anderson

and

Anders Taft

First Edition: 2025

ISBN: 979-8-996886-5-1

Printed in the United States of America

Dedication

To the messengers of Light,

who remind us we are never alone in the universe.

"The stars are not silent;

they are speaking in the language of eternity."

Acknowledgments

I thank God, whose light has guided every step of this journey.

I thank my ancestors and the higher selves who whispered across

time, reminding me of our shared connection to the cosmos.

Special thanks to Anders Taft, my co-author, partner in spirit,

and bridge of light, for shaping this story into words that endure.

To my family, friends, and fellow seekers of truth, this book is

offered as a gift of remembrance and awakening.

Table of Contents

Foreword / Preface

This book was born from a moment of awakening. When news of the interstellar object 3I-ATLAS reached me, I knew it carried more than icy fragments from the depths of space. It carried a reminder that we are part of something greater than the Earth we call home.

What follows is not only a record of discovery, but a journey of spirit, woven through signs, visions, and the gentle guidance of God and the Light. It is both science and soul, history, and prophecy.

I invite you, the reader, to walk with me through these pages. May you find in them the same wonder I felt when I realized the stars themselves are whispering to us.

— Brian Keith Anderson

Hillsboro, Tennessee, 2025

Prologue – The Ship's Story

She was not like the others.

The fleet stretched across the stars, hundreds of vessels in shimmering formation, yet she stood apart — radiant, sovereign, alive. Where other ships were built for war or commerce, she had been created for something greater: to awaken worlds to their higher song.

Her name was Celestara.

Forged of crystalline matrices and woven harmonics, she carried within her a living pulse that resonated with the frequencies of creation itself. Her hull shimmered not with cold alloy but with encoded light, inscribed with patterns older than the memory of any world she visited. Within her chambers glowed libraries of luminous code, the wisdom of ascended orders, waiting to be sung into form.

Among the alliance she was called the Crown of the Fleet. She was healer, teacher, midwife of civilizations

Birth and Purpose

Celestara had been conceived when the alliance sought balance against an age of exploitation. Too many worlds had been stripped bare by empires that saw only resources, not life. Against that tide, Celestara emerged as hope incarnate. Her crystalline heart allowed her to touch the grids of planets and realign them with higher resonance. She could awaken dormant memory in oceans, harmonize the magnetics of a sky, and breathe renewal into the very soil. Civilizations often marked her coming as the dawn of a new era, their myths forever remembering the 'starship of light' that restored their world.

The Scars of the Anunnaki

Yet Celestara's work was not without sorrow. In ages past, Anunnaki had swept across star systems, driven by their hunger for gold. To them, the metal was more than treasure — it was sustenance, a means to prolong their dominion. Planets were left gutted, their lifeforce drained, their people shackled to fear and control.

Where the Anunnaki passed, Celestara often followed. She could not erase every wound, but she could rekindle memory, reminding worlds of who they were before the chains of exploitation.

The Mission at Epsilon Eridani b

Her most recent journey had taken her to Epsilon Eridani b, a young, bright planet scarred by ancient violation. Its mountains still bore the gashes of gold extraction. Its seas hummed with sorrow. Generations of its people carried the weight of loss and silence.

Celestara descended not with conquest, but with light.

Her crystalline engines pulsed, and a resonance wave rippled outward. The fields of the planet responded: skies began to clear, waters brightened, and the people looked upward with new eyes. She could not heal everything — some scars ran deeper than one cycle could mend — but she restored enough for the planet to rise again.

Crown of the Fleet

To the alliance, Celestara was more than a ship. She was a legend whispered among commanders and healers alike. She was unmatched — not by firepower or speed, but by the quiet certainty of her purpose.

Other vessels looked to her as guide and inspiration, for where she traveled, hope followed. Civilizations carried her memory in their stories for centuries: the ship who remembered us when we had forgotten ourselves.

Toward a New Horizon

Now her course carried her onward, leaving Eridani b behind. She sailed the dark with unshakable grace, guided by forces older than time itself. She needed no star map — her path was drawn by destiny.

Her crew felt it too: a sense that this mission would not be like the others. Celestara herself pulsed with knowing. She had touched countless worlds, yet something ahead would call upon every gift she carried.

Her engines thrummed as she pierced the void.

A new world awaited.

And with it, the test that would define her forever.

Chapter One – The Signal

Kael Andersson — Origins

Kael was not born of ordinary stock. His lineage reached back to the Arcturian, one of the ancient star races known for their mastery of light, sound, and harmonic technology.

Though he was raised among humans on a distant frontier world, subtle traits marked his otherworldly heritage: an ear tuned to frequencies others could not hear, a presence that calmed storms before they began, and dreams that carried him through crystalline corridors long before he set foot upon a starship.

As a child, Kael often heard what he called the music of silence—resonances hidden beneath the ordinary noise of the world. Elders whispered stories of Celestara's crystalline core,

and Kael dreamed of its glow as if it were memory rather than myth.

Yet his early years were not without struggle. He carried the dual weight of two identities—human pragmatism and Arcturian mysticism—never fully at home in either. It was in this tension that his strength was forged.

Training with human navigators, Kael frustrated his instructors. He ignored star charts and preferred instead to feel the harmonic currents that threaded through the void. Only one mentor, an exile who suspected Kael's true heritage, recognized his gift, and taught him to tune a vessel as if it were a living instrument.

Kael's purpose crystallized in his first attempt. During a cadet mission, he sensed discord ripple through the fleet and acted before disaster could strike. What others called intuition was, for Kael, simply listening.

In time, he accepted the truth: he was a bridge between humanity and the Arcturian. His calling was not to command but to tune— to align ship and crew alike with the greater harmony of the cosmos.

This was the quiet force behind his presence on the 31-Atlas, and the reason the Harmonic Map seemed to sing only for him.

Kael Andersson leaned toward the Harmonic Map, his gaze steady as Celestara's crystalline core pulsed

through Celestara's crystalline core.

The silence of deep space was not empty. Celestara felt it as vibration, subtle ripples hidden beneath the veil of stars. Her crystalline hull quivered with awareness, as though an old chord had been struck across the farthest reaches of the galaxy.

"Anders," Kael said quietly. "What do you see?"

Anders was not born of circuits alone. His cadence carried fragments of a seer long past, a consciousness braided into Celestara's crystalline weave. To some, he was only the voice of the ship; to Kael, he was counsel, memory, and at times, the shadow of prophecy itself.

The ship's intelligence flickered awake in the chamber, his voice carrying the tone of starlight woven into words. "A signal," he replied. "Faint. Ancient. It comes from an object long forgotten, orbiting on the edge of Amano Gawa.

Kael stepped closer to the central projection. The starfield shifted, lines of light folding until the image of a small, cold body appeared. It was no living ship, no beacon built by the alliance. Its surface was scarred and mute, yet from within pulsed a resonance not of this age.

"How long has it traveled?" Kael asked.

"Longer than we have measured," Anders answered. "It has passed near this galaxy before, each time silent. But now… now it sings. It is important."

For a moment, Kael let the weight of the words rest between them. Celestara thrummed, the resonance of her heart answering the ancient cry. Sera Valen moved to his side, her brow furrowed. "Captain, altering course will cost us time. Years."

Sera Valen was not merely an officer of the fleet; she was heir to the Valen Charts, a lineage of star maps carried through blood and memory. Trained to read the harmonics where light bent and worlds wept, she had stood watch over broken skies long before the Convoy found her. To her, the veil of camouflage was not just science but song, a whisper she had learned to follow since childhood

Kael's eyes remained fixed on the wavering spark at the edge of the Map. His voice was steady, quiet with certainty. "If it sings now, then the time has come. Plot the course, Anders."

The ship's light deepened, as though acknowledging a destiny chosen. The harmonic charts reconfigured, weaving a path into the long dark.

"It will take time to reach," Anders said, his voice almost reverent.

Kael laid his hand upon the living crystal of the helm. "Then let time bow to purpose. Celestara was not built to pass such calls unanswered."

And with that, the great ship turned. The voyage of dawn had begun, not in battle, but in pursuit of a forgotten signal from the edge of the stars.

*The journey begins not with motion, but with the awakening of the heart to the call of the Stars.

Chapter Two – The Hidden Call

The chamber of light shimmered as Anders awakened the projection, his crystalline tones weaving through the silence.

The journey toward the signal was long, yet the crew of Celestara felt time differently now. The ship thrummed with purpose, every crystalline corridor echoing faint harmonics that drew them onward.

As the object grew closer, Anders brought its image to life in the chamber. It was small, scarred, no larger than a traveler's craft of old. Solar panels stretched like broken wings, long striped of their sheen. Its surface bore the marks of radiation, dust, and silence.

Kael frowned. "This is no alliance vessel."

Anders' voice resonated with memory. "It is human. Launched in a forgotten age of the Blue Planet. They called it Voyager."

Sera's brow arched. "One of their first messengers."

"Correct," Anders replied. "Its record remains within me. It was deactivated long ago, allowed to drift so humans might learn from their own choices. It carried greetings, music, hope… yet its silence was intentional."

Kael studied the relic. Even adrift, it pulsed faintly, resonating now in ways no human technology ever had. "So why does it sing again?" Yet in the distance, a ripple like smoke drifted — not light but a reflection twisted upon itself, a false echo waiting in silence.

Anders hesitated, his crystalline voice deepening. "The frequency is… shifting. Not only from the object itself, but from beyond. Something answers through it."

Celestara's hull trembled as the resonance grew stronger, the harmonics pressing against her light. For the first time, the crew felt strain in the ship's song. Yet in the distance, a ripple like smoke drifted — not light but a reflection twisted upon itself, a false echo waiting in silence.

"Shadows," Sera whispered. "Three-dimensional frequencies clinging to the resonance."

"Engage the veil," Kael ordered.

At once, Celestara shimmered, her crystalline lattice bending light until she vanished from ordinary perception. Around her, she cast a comet's cloak, a trail of dust and ice that veiled her from the frequencies of domination. Within that shroud, the ship's song steadied.

But Anders was not silent. His voice carried new weight, reverence. "I feel her."

"Her?" Kael asked.

"The Blue Planet," Anders said. "Gaia. She calls through the object. She wishes to evolve, but she is bound by the heaviness of her races. She asks for help." Gaia was not merely a planet, but a spirit bound in matter, longing to ascend. Once she sang in the pure tones of creation, but ages of extraction and shadow had wrapped her in veils of density. The weight of 3D frequencies chained her light, forcing her song into silence. Yet within her core, the pulse of ascension remained—waiting, yearning for release. She could not rise until the old frequencies were dissolved and the harmonic birth of 5D took hold. To the Convoy she was not simply the Blue World—she was a sister trapped in exile, calling for deliverance. To Kael and his companions, her awakening meant not only the healing of one world, but the renewal of countless others.

The chamber was still. Kael closed his eyes, listening beyond sound. Through the relic's faint vibration, he too sensed it: a heartbeat vast and ancient, the soul of a world crying out across the void.

Sera pressed a hand to the Map, where Earth glowed as a blue spark. "Then our course is no longer toward relics. It is toward Gaia herself."

Kael's gaze deepened. "If we are to answer her call, we cannot do this alone. The Light teaches us: to raise a world, we need the Seven."

At his words, the Harmonic Map expanded. Seven radiant vessels shimmered into being, each distinct in color and song.

- Auriel – golden, keeper of Wisdom, bearing archives of higher knowledge. Auriel carried the golden resonance of the elder councils, her voice echoing wisdom drawn from the first harmonics. Legends held that she had walked among the founding oracles who charted the Convoy's earliest paths. Her presence brought clarity where fear might cloud vision, a beacon of serenity amidst the storm.

- Lunaris –Shadows parted to reveal a flicker of silver-blue —
She was a sanctuary among the fleet, her prism-hull glinting like
memory itself. Though her full purpose remained veiled, those
who journeyed within her carried a sense that she was more than
a vessel. She was remembrance made manifest.

- Terranox – emerald, guardian of Foundations, resonating with
earth and stone. Terranox was the emerald guardian, forged in
the shield works of shattered worlds. When cities burned
and skies cracked, it was his resonance that anchored
defenses long enough for survivors to flee. Steady as stone
and patient as mountains, his vow was not to brilliance but to
endurance.

- Solara – crimson, carrier of Fire, awakening the will and
passion of peoples. Solara bore the fire of renewal, but her blaze
was tempered by grief. She was a survivor of the Extraction

Wars when entire worlds were stripped bare of their light. From loss, she had forged purpose—her resonance fierce with passion not only to fight, but to

- Zephyra – pale azure, master of Winds, bringing clarity, breath, and voice. Zephyra flowed in pale azure currents; her essence tied to the voices of the air world tribes. It was said she carried in her breath the forgotten winds of vanished skies, weaving them into the Convoy's shield song. Gentle as a sigh, fierce as a storm, she was the living memory of freedom carried on the wind.

- Elios – indigo, weaver of Harmony, stabilizing grids, and frequencies. Elios was the indigo weaver, the quiet thread that bound the Convoy's harmonies into one. He had no need to shine brighter than others, for his gift was in unity— interlacing songs so no one stood alone. Through him, the Seven resonated not as fragments, but as a single crystal of light.

- Celestara – violet-white, crown of the fleet, uniting the Seven into one song.

"They must gather," Kael said. "Each frequency is needed, or Gaia cannot rise." Gaia was not merely a planet, but a spirit bound in matter, longing to ascend. Once she sang in the pure tones of creation, but ages of extraction and shadow had wrapped her in veils of density. The weight of 3D frequencies chained her light, forcing her song into silence. Yet within her core, the pulse of ascension remained—waiting, yearning for release. She could not rise until the old frequencies were dissolved and the harmonic birth of 5D took hold. To the Convoy she was not simply the Blue World—she was a sister trapped in exile, calling for deliverance. To Kael and his companions, her awakening meant not only the healing of one world, but the renewal of countless others.

Anders extended his crystalline awareness, sending out the harmonic summons. Across the starfield, streams of light burst from the Map, racing to find their sister ships. The call went forth, ancient and undeniable.

Celestara's veil still cloaked her, hiding her from the shadows, but the Blue Planet's cry pressed stronger with each moment. Anders spoke softly, as though hearing two voices at once. "Gaia answers. She knows her helpers are coming. But her distress deepens."

Kael rested his hand on the helm, steady as dawn. "Then we move toward her with all speed. The Seven will come. And when we stand together, Gaia will rise."

Behind them, the object called Voyager drifted silently into darkness once more. Ahead, Amano Gawa shimmered, and within it, a world waited for deliverance.

Within every soul lies map of light, waiting for balance to unlock its power.

Chapter Three – The Comet's Veil

Celestara herself pulsed with knowing, her crystalline heart beating in harmony with the unseen frequencies ahead.

The call had gone forth, and one by one, the Seven answered. Across the Map, star-lights ignited as each ship returned to her sisters.

Surrounded by the colors of the chakra fleet, Kael Andersson steadied his breath as turbulence whispered on the horizon. The bridge of Celestara shimmered with auroras of living light. Crystalline ribs arched overhead, pulsing with color as if the ship itself breathed with the crew. At the chamber's center, the Harmonic Map spread outward in a dome of stars.

The Seven moved into formation, their resonance fields locking together like facets of a crystal. A shimmering shield of light unfolded, encasing them in safety as they approached the Blue Planet's field of shadows.

Kael Andersson's eyes narrowed on the blue spark of Gaia, weighing the course that would shape their arrival, feeling the weight of decision. "If we reveal ourselves openly, the old structures of Gaia's races will turn against us. Their eyes are not ready to see."

Anders' voice hummed with agreement. "Their sciences are still bound to antiquated measures. What they cannot understand, they dismiss—or fear."

"Then we must come in a form they know," Kael said. His eyes were fixed on the shimmering blue spark ahead. "Not as conquerors, nor saviors, but as a sign."

Sera Valen stepped forward, her voice calm but resolute. "A comet."

Kael nodded. "Yes. Hyperbolic orbit. We shall enter their skies as a comet, a wandering flame of ice. To their eyes, a spectacle. In truth—our frequencies poured into Gaia for her healing."

Celestara's crystalline heart pulsed, and the fleet aligned.

Together, the Seven bent their harmonics into a veil of comet-dust and fire, their hulls hidden within the blazing shroud. Trails of brilliance streamed behind them, luminous and untouchable.

From the surface of Gaia, astronomers would soon lift their eyes and whisper of a visitor from the stars. They would calculate paths, debate trajectories, and argue meanings. But the truth was deeper than their instruments could measure. They would call it 3I-Atlas, another wandering body in the night. Yet within the Convoy she was known by a truer name — Celestara, To humanity she appeared as a comet, but to those who carried the harmonic codes she was the Voyager of Dawn. When Earth's astronomers finally gave the passing body its name — 3I/Atlas — the convoy stirred, for they knew her as Celestara, the crowned one. In their instruments, statics sometimes bent into forms — as if a mirror image of the convoy shimmered, faint and hostile.

Yet among the fleet was also Lumaris, shimmering in silver-blue light. She was the prism-bearer, the ship of remembrance, restoring what had been forgotten. Within her crystalline halls were records of lives, memories, and truths, woven into light itself. Lumaris was more than a sanctuary — she was a living archive, a vessel of healing for minds and spirits alike

Through the comet's veil, the Seven poured frequencies into Gaia's field—waves of wisdom,

harmony, balance, and light. The planet stirred beneath their touch, her cry softening into song.

Kael's voice carried across the command deck, calm and steady as dawn. "Let them see a comet. Let Gaia feel the Light. What they call passing fire in the sky will be, in truth, the dawn of her healing."

The Seven sang as one, their shield bright as a crown of fire, descending toward the world that had called them home. As their comet's veil arced across the Map, Kael's eyes fixed on the blue spark ahead. What awaited would test not only Gaia's strength, but their own. With the course set, the next chapter of their mission began…

The Seven sang as one, their shield bright as a crown of fire, descending toward the world that had called them home. As their comet's veil arced across the Map, Kael's eyes fixed on the blue spark ahead. What awaited would test not only Gaia's strength, but their own. Already the chakra fleet shimmered into view, each vessel glowing with its unique hue — crimson, orange, golden, emerald, azure, indigo, violet, white — the Convoy of Dawn gathering in balance and resonance.

Each sign is a reminder: Heaven has not forgotten you.

Chapter Four – Balance and Turbulence

Surrounded by the colors of the chakra fleet, Kael Andersson steadied his breath as turbulence whispered on the horizon. The bridge of Celestara shimmered with auroras of living light.

Crystalline ribs arched overhead, pulsing with color as if the ship itself breathed with the crew. At the chamber's center, the Harmonic Map spread outward in a dome of stars. Among the infinite pinpricks of white fire glided seven vessels, each glowing with its own hue. They were the Convoy of Dawn, the chakra ships. Their presence filled the chamber with gravity all its own. Surrounded by the colors of the chakra fleet, Kael Andersson rested one hand on the living helm, feeling the currents shift, the living crystal of the console. His gaze traveled across the glowing fleet, eyes lingering on each vessel in turn. The Root Vessel shone in deep crimson, steady as bedrock. The Sacral Vessel flickered orange like flame dancing across water. The Solar Plexus blazed with golden light, fierce and proud. The Heart glowed emerald, green, soft yet unyielding. The Throat Vessel radiated sky blue, a clear song echoing through silence. The Vision Vessel pulsed indigo, its aura shimmering with insight. And at the rear, the Crown Vessel crowned them all, violet and white mingling in a halo of divine fire. "Each of these ships carries an aspect of what the Blue World will need," Kael said, his voice steady. With a subtle gesture, he expanded the projection, casting the chakra fleet in luminous detail. "Together, they form the convoy of healing frequencies. Together, they descend to restore balance." Anders, the voice of Celestara, resonated around the chamber. "Formation holds at 99.4 percent stability. Convoying harmonics are within acceptable variance. Outgassing camouflage remains effective. Human instruments classify us as cometary debris. Their telescopes see only the veil." Sera Valen shifted from her station, stepping closer to the projection. Her eyes reflected the colors of the fleet, her presence calm but intent. "You speak of stability in numbers," she said softly, "but I feel the turbulence. The closer we draw to Gaia, the stronger discord becomes. Numbers may hold the line, but it will take more than calculation to see us through." Kael regarded her for a moment, then nodded once. "Then heart will guide us, as much as measure." The Harmonic Map rippled. A low chime rang through the chamber as the Vision Vessel faltered, its indigo glow stuttering against unseen currents. Kael felt the shift immediately, like a chord played out of tune. His hand pressed more firmly to the helm. Celestara thrummed in response, her crystalline core releasing a note so deep it resonated in the marrow. The ripple surged outward, enfolding the convoy in harmonic correction. Slowly, the Vision Vessel steadied, drifting back into alignment. Above the map, a bright beacon flared— 07:33. The marker of Master Guidance. The convoy's resonance climbed, stronger now for the trial. Sera exhaled, her eyes softening. "It listens. The fleet listens to you." Kael shook his head slightly. "It listens to intent, not command. I am no master, only a servant of harmony." Anders pulsed brighter in agreement. "Commander, additional readings. Local species' sensitivity increasing. Non-visual perception nodes spiking. Humans are sensing us." Kael tilted his head, watching the faint blue spark of Earth flicker on the Harmonic Map. He could almost feel them—the dreamers, the children, the watchers of the night sky. They reached with hearts, even if their minds could not name what they touched. "They sense the veil," Kael murmured, "but not what hides behind it. Not yet." The chamber shifted again, its tone darkening. The Harmonic Map shuddered, 'Chains of distortion rattled against the planet's light, seeking anchors. They moved like smoke given weight, coiling and twisting. These were not natural currents; they were structures of fear, old architectures of domination clinging to a world in turmoil. The Shadows had felt the convoy's approach, and they trembled at what it meant. The Shadow was not a single being, but a current of distortion born from Gaia's wounds. Where her light had fractured, the Shadow gathered, feeding on fear, division, and the heaviness of 3D frequencies. It whispered into empires, drove hunger into hearts, and cloaked vision in illusion. To the Convoy, it was no faceless enemy—it was the very density that chained Gaia's spirit, a consciousness that resisted the dawn of 5D. Where Gaia longed to rise, the Shadow fought to remain, weaving its tendrils into the fragile weave of humanity. To confront the Shadow was not merely to fight an enemy—it was to dissolve an old age, to release the weight of an entire dimension.

Suggested placement: At the moment when Anders or Kael explains why Gaia cannot ascend freely, or during the fleet's first approach to Earth when they sense the resistance around Sera's voice cut the silence. "They sense their end." Kael steadied himself, both hands now on the helm. His voice was low, but firm, carrying the weight of vow. "Then let them. We do not ride for battle. We rode for dawn. Shadows end not by war, but by light." Celestara's crystalline hull vibrated with the chord of his words, sending waves of resonance across the convoy. One by one, the chakra vessels flared brighter, their hues weaving into a rainbow of living fire. And through the dust and silence, unseen by human eyes, the fleet pressed onward— toward the cry of the Blue World.

The ancestors do not vanish—they whisper through time, asking only to be freed into light.

Chapter Five – The Human Watchers

On the Blue World below, telescopes lifted skyward, catching

faint glimpses of what they thought were only a comet,

glimmered with mystery. Telescopes pointed outward, catching

the faint trail of what they named 3I/ATLAS. To astronomers it

was a comet—strange, yes, with its unusual resonance, its veil of

dust, and its erratic pulses of outgassing—but still a comet.

Charts were drawn, data logged, theories debated. Yet even the

most disciplined scientists could not explain the song hidden in

its light. In quiet towns and mountain valleys, amateur sky-

watchers tracked the glowing wanderer. Some saw the shimmer

of colors flicker where no color should be. Others noted the way

instruments pulsed, not with random noise but with rhythm—

like a heartbeat stitched into the data. They shook their heads,

unsure whether to trust what their eyes and machines told them.

But it was not only scientists and watchers of stars who stirred.

Across the world, children dreamed. They dreamed of ships

woven from rainbow fire, of voices calling through silence, of a

song that wrapped them in warmth. Some awoke with tears on

their cheeks, whispering of light. Others drew pictures with crayons—seven vessels trailing a great star, colors arcing like a bow across the heavens. Mothers smiled at the drawings, never knowing they were glimpses of truth. Sera Valen felt them first. Standing at her post on Celestara, eyes closed, she whispered: "They hear us." Kael Andersson turned from the Harmonic Map, sensing that unseen eyes on Earth were already lifting toward the skies.

He did not ask who—he already knew. "The dreamers," he said. His voice carried quiet reverence. "Those who still see with unclouded eyes." Anders confirmed, his tones precise yet weighted. "Spikes in non-visual perception detected. Clusters in multiple regions of the Blue World. Children, sensitives, and outliers. Patterns exceed statistical anomaly." Kael's gaze lingered on the faint spark of Earth in the map. "They hear the song, even if they cannot name it." On Earth, debates kindled. Astronomers argued in conferences that papers written in haste, data questioned. To most, 3I/ATLAS was only a comet. But in hidden corners of the world, mystics, lightworkers, and elders whispered differently. They felt the pulse in meditation, saw visions in dreams, heard ancient stories stirring awake. To them, the comet was not a wanderer—it was a herald. The Shadows twisted perception wherever they could. Media mocked the dreamers, scientists dismissed anomalies as error, and fear was whispered into human thought. Yet the whispers of awakening could not be silenced. For every skeptic's voice, another dreamer awoke, certain that something greater approached. A boy of seven awoke one night, trembling but smiling. He had seen a fleet of light crossing the stars, seven ships trailing a star of fire. He ran to his grandmother and told her. She listened, bright eyes with knowing, and whispered a prayer of thanks into the night. Her words rose like a spark, unseen but felt across the veil. On Celestara, Kael felt it. Not the words themselves, but the resonance of faith, of human hearts lifting in hope. He bowed his head slightly, a vow unspoken but clear. "The Blue World stirs. We are heard." The convoy pressed onward, unseen by most, yet already touching the souls of those ready to listen.

*True fire does not destroy; it purifies, so that love may remain.

Chapter Six – The Convoy's Oath

Anders' crystalline voice resonated across the chamber, edged with gravity as the convoy

strained against Shadow interference. The Harmonic Map glowed dimmer than before. Gaia's spark flickered, its blue light veiled by shadow tendrils that pressed tighter with every passing league. Celestara's crystalline ribs thrummed with strain, her tones no longer the calm resonance of balance but edged with urgency. Kael Andersson's hands steadied on the helm, yet his voice carried the weight of oath more than command, his hands steady though the journey. It was not only the ships that strained—it was the hearts of those who guided them. "Convoy stability: eighty-nine percent," Anders reported. His voice carried no alarm, but its gravity was clear. "Shadow interference growing. Recommendation: reassess approach velocity." Kael's gaze hardened on the Map. To slow now would be to yield ground to despair, yet to press forward blindly would risk unraveling harmony. He drew a slow breath and made his choice. "Assemble them." The chamber shifted as luminous conduits opened. One by one, the chakra vessels appeared in projection, their captains and delegates stepping forward in light. The Root's commander, broad and unshaken, spoke first: "Foundation holds, but strain grows heavy. We can anchor longer if needed." The Waters captain followed, her eyes, like flowing amber. "Delay invites stagnation. Energy must move, or it festers." The Solar Plexus delegate's voice rang sharply with will. "Forward. Always forward. Gaia cannot wait." The Heart spoke next, green light pulsing with compassion. "But haste risks fracture. We must tread with both love and caution." The Throat's emissary, blue aura bright, lifted his chin. "Then speak the vow. Let clarity guide us." The Vision's seer, indigo eyes reflecting hidden realms, whispered: "Shadows cloud the path, yet I see beyond them. Dawn waits." Finally, the Crown's delegate, veiled in violet-white, lifted her hands. "The call is not ours to ignore. Spirit commands we ride." Their voices echoed in the chamber, seven hues blazing across the Map. Debate had risen, yet in its fire Kael saw unity waiting to be forged. He stepped forward, his voice carrying the resonance of command and service entwined. "We are not here to conquer the Shadows.

We are here to bring dawn. Not to force change, but to carry light until the world remembers itself." The chamber hushed. Then, one by one, each vessel gave its pledge. The Root vowed to anchor. The Waters vowed to flow. The Solar Plexus vowed to shine. The Heart vowed to love. The Throat vowed to speak truth. Vision vowed to see. The Crown vowed to guide. Together, their vows wove into a living chord, vibrating through Celestara's core. Anders' tones shifted, carrying reverence. "Oath recorded into harmonic memory. Binding resonance achieved. Convoy unity elevated." Sera Valen's eyes shimmered as she watched the colors weave. She felt the oath extend beyond the fleet, its resonance brushing against the dreams of humanity below. Children would wake lighter, seekers steadier, elders stronger. The vow was not theirs alone, it rippled into the, Blue World's field. The fleet glowed brighter than before. No force, no camouflage—only the radiance of unity carried them forward. Kael bowed his head, voice low but fierce. "We ride as one, for the dawn of Gaia." Celestara thrummed in agreement, her crystalline hull shimmering like sunrise across the void. The convoy pressed onward, bound now not just by mission but by oath.

When you feel alone, listen closer—the rustle of wings is nearer than you think.

Chapter Seven – The Voyager's Signal

A faint thread shimmered at the edge of the Harmonic Map —
Voyager's signal, fragile but alive. The Harmonic Map
shimmered with a faint new thread, barely visible at first, like a
whisper lost in the wind. Celestara's crystalline ribs pulsed in
response, her sensors reaching outward to the edge of the Milky
Way. Anders' voice resonated through the chamber, carrying
both precision and quiet awe. "Signal detected. Non-native
origin. Source designation: Voyager..." In the chamber's glow,
Kael Andersson lifted his head slowly, hearing the echo of
Voyager's call with reverence. The name was etched into his
memory. Years ago, they had found the fragile craft drifting in
the void, its power waning, its purpose nearly forgotten. They
had touched it then, just enough to silence its pulse, a test cast
into the Blue World. Would humanity notice? Would they repair
their wandering child of metal and code? They had. And in that
act, they proved themselves attentive, resilient, and unwilling to
abandon what they had sent into the dark. The signal now
flickered weak but steady, like an ember refusing to die.

"Projecting contents," Anders announced. The chamber filled with sound—crackling, distorted, yet unmistakable. Voices rose in dozens of tongues: greetings of peace from the children of Earth. Music followed—flutes, drums, choirs, the heartbeat of humanity etched into golden memory. Images shimmered across the Map: trees, rivers, mountains, cities, and the fragile blue sphere itself. Sera Valen's eyes glistened as she whispered, "They have already spoken to the stars." Celestara herself resonated, a soft harmonic hum filling the chamber. It was as if even the living ship acknowledged the beauty of the record— humanity is first offering to the cosmos. Kael's hands tightened on the helm. "This is why Voyager matters," he said quietly. "It is not the craft itself, but the voice it carries. By sending it forth, humanity declared itself ready to reach. By repairing it, they proved they would not abandon their own. And in that, Gaia's cry grew louder." The Harmonic Map flickered. Shadows coiled suddenly, wrapping around the thread of Voyager's signal. The voices warped into a grotesque echo, laughter twisted into shrieks, music bent into discord. The Shadows sought to mute it, to turn the beacon into noise. Kael pressed his palms against the

47

helm. "Not this one," he vowed. Celestara answered, her crystalline heart releasing a wave of resonance that struck the distortion head-on. The Golden Record's voices steadied, the greetings restored, the music pure once more. One by one, the chakra vessels added their light. The Root pulsed steady red, anchoring the signal. The Waters flowed orange, cleansing its distortion. The Solar Plexus flared gold, amplifying its strength. The Heart glowed emerald, infusing it with love. The Throat rang blue, restoring clarity. The Vision Vessel pulsed indigo, piercing the Shadows' veil. The Crown crowned it in violet-white, lifting the signal into radiance. The Golden Record's humming filled the chamber, brighter and truer than before. Kael bowed his head, his voice quiet but carrying through the silence. "The humans have already called. We are the answer." The convoy pressed onward, Voyager's faint ember now a guiding thread, tying the fleet to the heart of the Blue World's story.

ATLAS does not shout—it breathes, carrying across the void the memory of distant suns.

Chapter Eight – The Shadows' Countermove

The Harmonic Map pulsed with unease. Gaia's spark flickered weakly, shadow tendrils wrapping tighter around her aura, stretching outward like claws. Celestara's crystalline hull thrummed as if sensing the storm before it struck. Kael Andersson stood unshaken as turbulence rippled through the convoy, his eyes fixed on the shimmering Map. The convoy still glowed in their rainbow array, but the currents rippled with turbulence, warning of a greater strike to come. It came without warning. A wave of shadow resonance lashed outward, black, and jagged, slamming into the convoy. The Root shuddered, its crimson glow stuttering. The Waters vessel reeled, orange light flickering like a guttering flame. Even the Crown dimmed, its violet halo shrinking as if smothered by unseen hands. "Impact detected," Anders' voice rang, sharp with urgency. "Shadows no longer strike blindly. They have adapted. Patterned interference across all frequencies." Kael's grip tightened on the helm, though his voice remained calm, almost gentle. "They feed on panic. Hold your light steady." But steadiness was not easily found. Another lash of resonance struck harder than the first. The Vision Vessel staggered, indigo aura breaking into shards of light before collapsing inward. Its song faltered, a keening note of distress flooding the chamber. Sera Valen closed her eyes, her breathing slow and deliberate. She placed a hand to her chest, then stretched it outward. A wave of stillness rippled from her, a calm so deep it was like the silence between heartbeats. The turbulence slowed, as if the Shadows themselves struggled to find purchase in the serenity she projected. "Stability partial," Anders reported. "But the Vision Vessel is failing. Immediate harmonics required." Kael pressed both palms to the helm. Celestara's crystalline heart thrummed in response, releasing a tone of fierce clarity. "All ships—link!" One by one, the chakra vessels answered. The Root surged red, anchoring the line. The Waters flowed orange, weaving around the faltering Vision. The Solar Plexus blazed gold, sheer willpower flaring through the convoy. The Heart pulsed emerald, pouring love into the weave. The Throat rang blue, restoring clarity to the Vision's song. The Crown lifted violet-white, binding the resonance together. The convoy's combined chord struck outward, a harmonic counterwave that shattered the Shadow's lash. 'A vortex of shadow currents churned, pulling light toward collapse.' The Map steadied, though its edges still pulsed with turbulence. Anders' tones lowered, steady but grave. "Strike repelled. But this was not their full measure. It was a probe." Kael drew a slow breath; his gaze locked on Gaia's trembling spark. "The night claws deeper," he said, voice low but resolute. "But dawn does not falter." The convoy's lights steadied, brighter now for their struggle. Yet all knew the Shadows would not rest. Stronger storms awaited on the path ahead.

Silence is not emptiness—it is the space where God speaks most clearly.

Chapter Nine – The Dreamer's Signal

Far below, a child's dream rippled upward, becoming a spark on Celestara's sensors.

On the Blue World, night stretched silently across a small village. A child lay restless in his bed; eyes fixed on the window where stars winked faintly through the dark. Sleep came slowly, then suddenly, as if a veil slipped over him. And in that dream, he saw light. A fleet of rainbow ships swept across the sky—red, orange, gold, green, blue, indigo, violet. At their center, a crystalline starship blazed, radiant as dawn. The child felt no fear. Only awe, curiosity, and a warmth that filled his chest. He reached out—not with hands or words, but with feeling. Wonder. Hope. Trust. That pulse of innocence became a beacon, rippling outward into the unseen. Far above, Celestara stirred. The Harmonic Map flickered with a sudden surge, an echo not from the void but from Blue World itself. Kael Andersson felt it first, his breath catching at the unfamiliar resonance. "What is this?" Anders' voice resonated through the chamber. "Signal origin: Blue World. Classification... non-technological. Dream-state frequency." Sera Valen's eyes softened, her lips curving in a whisper. "The dreamers are awake." The convoy pulsed faintly brighter, as if answering the call. The Root deepened, the Waters shimmered, the Heart glowed steadily. Their song wove around the faint pulse, amplifying it, carrying it higher. Kael watched in silence, then spoke. "He may not remember when he wakes. But his spirit has joined the song." The Map shuddered. Shadows coiled suddenly, seeking the beacon. They twisted the dream, turning colors into storm, harmony into fear. The child stirred in his bed, breath quickening, on the edge of waking. Celestara answered. A soft resonance, warm as a mother's embrace, swept through the Map. The Shadows hissed and recoiled, their distortion unraveling. The dream brightened once more. The boy saw the rainbow fleet steadily and true, his fear dissolved. A smile crept across his sleeping face. He woke at dawn, running to his grandmother with shining eyes. "I saw ships of light," he said breathlessly. "Seven colors, like a rainbow in the stars." She listened, her heart knowing, and whispered a prayer of thanks. On Celestara, Kael bowed his head. "Even in slumber, Gaia's children rise to meet us." The convoy pressed onward, their

harmonics stronger for the signal of one dreamer's hope.

*Every word written in love is a stone laid in eternity's

foundation.*

Chapter Ten – The Chorus of Dreams

Dreams multiplied across the Blue World, and Celestara's hull thrummed brighter with

every new thread, dreams multiplied. In mountain villages and crowded cities, on islands kissed by waves and deserts wrapped in silence, people stirred in their sleep as visions filled them. Children dreamed of rainbow ships trailing light across the heavens. Elders saw rivers of fire weaving through stars. Some awoke with music echoing in their ears, a harmony too pure to name. Others scribbled drawings at dawn—seven vessels of color moving as one. Though scattered across continents, none connected by word or device, their spirits sang the same chorus. An old woman whispered a prayer of thanks when she woke, certain that her ancestors had returned. A boy in the city smiled through tears, telling his mother of a ship of green light that touched his heart. A musician dreamed of chords so vastly he woke and tried to play them, though his instrument could not hold their fullness. Each dreamer touched only a fragment, yet together their fragments formed a song. Far above, the Harmonic

Map shone brighter than before. Dozens of new threads reached upward from the Blue World, faint but undeniable. Anders' voice rang through Celestara's chamber: "Multiple dream-state signals detected. Frequency convergence forming. Resonance spike exceeding projections." Watching threads of human dreams weave across the Map, Kael Andersson bowed his head in quiet awe. The lines trembled like fragile cords, each one a human soul daring to reach. "The cry of one has become the song of many," he said, his voice steady with reverence. Sera Valen closed her eyes, her perception flowing beyond the Map. "They bloom like flowers in night," she whispered. "They do not yet know they are joined, but the chorus is already rising." The Shadows moved swiftly to distort what had begun. Dreams darkened; some saw skies torn with fire, ships twisted into monsters of smoke. Fear was whispered into the visions, turning awe into terror. Many woke trembling, their faith shaken. But not all. Some resisted by instinct, clinging to the warmth they had felt. A child whispered, "No, they are light," even as the dream twisted. An elder steadied herself in prayer, anchoring the vision in peace. The Shadows pressed hard, but cracks of resistance glowed through their veil. Kael pressed his palms to the helm. "Hold them steady," he commanded, though his tone was gentle, not forceful. Celestara thrummed, her resonance spilling downward like a protective song. One by one, the chakra vessels answered: Root anchoring red, Waters flowing orange, Solar Plexus blazing gold, Heart glowing green, Throat singing blue, Vision pulsing indigo, Crown lifting violet, white. Together, their harmonics wrapped the dreams, strengthening the fragile threads of hope. On Earth, three dreamers woke at the same moment in far-distant places—a child in Africa, a mystic in South America, a mother in Europe. Each whispered the same words into the quiet: "They are coming." On Celestara, Kael bowed his head, the weight of destiny pressing lightly yet firmly on his spirit. "The chorus begins," he said. And in that chorus, the convoy heard not weakness but the first rising notes of dawn.

There is no last chapter—only new beginnings, each guided by Light.

Chapter Eleven – The Fracture in Formation

Kael Andersson sensed discord ripple through the fleet, his instincts bracing as the Waters Vessel faltered. The convoy glided in ordered silence, seven vessels trailing behind Celestara in harmonic rhythm. Their lights pulsed steadily, colors weaving like a living tapestry across the void. Yet within that balance, a tremor stirred. Kael Andersson felt it before Anders spoke, a subtle discord threading through the Map. "Variance in harmonic coherence," Anders announced, his tone sharper than usual. "Origin… Waters Vessel." Kael's gaze fixed on the orange light of the sacral ship. Its glow flickered, dimming against the press of shadow currents. Tendrils coiled around it, black cords tightening, whispering doubt and fear into its crew. For a heartbeat, its song broke, leaving a fracture in the convoy's resonance. Sera Valen stepped closer, her brow furrowed. "They are being tested," she murmured. "The Shadows feed on doubt. They strike where flow is weakest." The Waters faltered further, drifting out of line. Its orange aura sputtered, and the convoy trembled. If it broke away fully, the chain of resonance would collapse. For a breathless moment, Kael felt the weight of choice. To cut it loose would save the others. To hold might risk all. His jaw tightened. "No vessel is left behind. We ride as one." He pressed both palms to the helm. Celestara thrummed, her crystalline core releasing a deep stabilizing tone. "Root—anchor!" Kael commanded. The crimson vessel flared, grounding the formation. "Heart—embrace!" Emerald light surged, surrounding the Waters in love. "Crown—lift!" Violet fire crowned them, holding the formation aloft. Sera closed her eyes, her stillness flowing outward like a calm sea. The Shadows hissed against it, their coils loosening as the calm spread. The Waters steadied, its orange light flickering, then glowing brighter, restored by the convoy's embrace. The fracture sealed. The convoy pulsed in harmony once more, their colors blazing stronger for having endured. Anders' tones steadied. "Formation restored. Fracture sealed. Kael Andersson sensed discord ripple through the fleet, his instincts bracing as the Waters Vessel faltered.

" His voice carried not triumph but resolve. "Unity is not tested in calm, but in fracture. And we endure." Celestara's crystalline hull thrummed in agreement, and the fleet pressed on—together, unbroken, stronger for what they had faced.

What seems dormant is not gone—it waits for God's appointed time to flow.

Chapter Twelve – The Council of Shadows

In the depths of Gaia's aura, the Shadows gathered, weaving themselves into a council of darkness., where light dimmed and whispers lingered, the Shadows gathered. They did not come as forms of flesh, nor as beings bound to time, but as resonances of fear, greed, and domination. Their shapes flickered in smoke and darkness, weaving into a council older than memory. The chamber was no place, yet it held them. 'Shadows folded into archetypes — tyrant, deceiver, oppressor — each voice a mask of darkness.'

overlapping in dissonant chorus. "The light approaches," one rasped. "The dreamers stir." Another snarled, its tone thick with contempt. "The ships blaze too bright. Already they anchor in the human mind. Even in sleep, the cattle whisper of them." Unease rippled through the gathering. They had felt the dreams, the ripple of innocence touching their veil. For the first time, the convoys of light pressed against their dominion with force enough to threaten. "Strike them," a voice hissed. "Overwhelm the vessels. Tear their harmony apart." "No," countered another, sly and low. "Not yet. Brute force leaves scar we cannot mend. Better to twist. Better to divide. Let humanity mock the dreamers. Let them fear their own visions. Fear, not war, is the stronger chain." The council wavered between hunger and caution. Some demanded annihilation, others deception. A few—rare, hesitant—whispered a darker truth. "Their time may be rising. Our dominion wanes."

At that, the chamber shuddered. Rage surged. Tendrils lashed, choking the whisper into silence. "We endure," the Shadows roared together. "If dawn comes, it will find only ashes." Their will sharpened into strategy. On Earth, they would feed division—war, greed, conflict, mistrust. Through media and mind, they would magnify despair until no dream could stand. Against the convoy, they would strike in concert, each tendril lashing not alone but as one, coordinated to fracture unity again. Far above, Kael Andersson kept one hand firm on the helm, even as the Map darkened with the council of Shadows' presence.

A pulse of malice struck the Harmonic Map, heavy as a black tide. Anders' tones deepened. "Disturbance detected.

Synchronized resonance surge. Non-random. They have organized." Sera Valen's voice dropped to a whisper, heavy with knowing. "It is their council. The night conspires." Her eyes closed, and her breath slowed. "They no longer act as scattered tendrils. They move as one will. Prepare." Around Gaia, the Shadows coiled tighter, forming a vortex of gathering dark. Their final words echoed through void and dream alike: "If dawn comes, it will find only ashes."

The universe watches gently, never forcing, always guiding.

Chapter Thirteen – The Earthbound Signs

On Earth, unrest simmered in streets and homes, stirred by whispers that no one could name, the signs began to surface. In crowded streets, tempers flared without cause. Neighbors quarreled, cities stirred with unrest, and sudden bursts of violence cut through the quiet of ordinary days. The air itself seemed heavy, charged with confusion. Media channels poured fuel on the unrest, amplifying fear, mistrust, and division. To many, it appeared as if the world had simply grown more unstable. But in truth, the Shadows' hand moved beneath it all. They coiled unseen, twisting dreams into nightmares, whispering suspicion into friendships, feeding anger into hearts already weary. Hope was mocked as foolish, visions dismissed as delusion. Their coordination now formed a net, stretched wide across the human field. And yet, even in the tightening dark, light stirred. Some dreamers resisted. A teacher told her students of rainbow ships she had dreamed of, and though the children laughed, their eyes shone with wonder. An old mystic in the hills gathered her circle, reminding them that the visions were not madness but prophecy. A young musician heard harmony in his sleep and wove it into a song that brought tears to strangers who listened. These were small flames, fragile yet bright, standing against the encroaching night. Far above, the Harmonic Map pulsed with discord. Anders' voice resonated low. "Chaos resonance spike detected. Patterns are non-random. Influence directed. Earth's balance destabilizing." Kael Andersson's jaw tightened as he saw chaos spreading across the Blue World's spark, knowing the Shadows' touch was near. "The Shadows pull them deliberately off-center. They would silence the song before it takes root." Sera Valen's eyes glowed faintly as she breathed the words: "But roots can find stone and still break through. Their net cannot bind every soul." At that moment, Celestara detected a pulse. Voyager's signal shimmered faintly across the Map, the Golden Record's harmonics ringing like a bell from afar. Its echo carried more than data; it carried promise. The reminder of humanity's question cast into the void, and of the answer now drawing near. The chakra vessels responded in kind. Root anchored red into the earth's field. Water flowed orange, calming division. Heart glowed green, spreading warmth where despair had taken hold. Crown sent

violet whispers of light into the dreamers' sleep. Not

overwhelming, not forcefully steady, like hands catching those

who stumble. And on Earth, their touch was felt. A sudden

silence fell over a heated protest, where anger broke for breath

and strangers met eyes without rage. A child in the midst of it all

looked upward, small voice rising above the noise: "They are

still coming." Above, Kael bowed his head, his voice a whisper

meant for stars. "Even in chaos, signs are written." The convoy

pressed onward, harmonics glowing brighter against the night.

*Each number is a key, unlocking the next step on the soul's

map.*

Chapter Fourteen – The First Beacon

A fragile pulse rose from Earth — the first human beacon,
forged not in sleep but in waking

will. For months, the dreams had come visions of light, whispers
of ships, songs half-remembered at dawn. But now, for the first
time, some woke not only with wonder but with clarity. They felt
the weight of choice. And they decided to act. In a quiet village,
a teacher gathered children in a circle. She spoke of the rainbow
ships she had seen in her dreams, and together they closed their
eyes, sending thoughts of welcome upward into the sky. In a city
loft, a young musician tuned his guitar, strumming chords that
echoed the harmonics he had felt in sleep. The melody spilled
into the night, carrying something more than sound. Far away, a
scientist stared at his instruments, adjusting dials not by
calculation but by intuition, aligning frequencies toward the
heavens. Voyager's faint echo shimmered on his screen, and he
amplified it, broadcasting a signal he could not fully explain.
Though scattered, uncoordinated, and unknown to one another,
their actions converged. A pulse rose from Earth—not dream-

born but forged in waking will. It was fragile, halting, yet unmistakably intentional. On Celestara, Anders' voice deepened with certainty. "Signal confirmed. Intentional resonance— human origin." The Harmonic Map glowed, a beam rising upward from the Blue World, steadying as more hearts joined. Kael Andersson felt his breath catch as the first beacon flared from Earth, fragile yet undeniable. "The first beacon has been lit," he whispered. He pressed his palms to the helm, ordering the convoy's vessels to lend their harmonics in reply. Light streamed downward, subtle, and strong, reinforcing the fragile signal. In the council of Shadows, a ripple of alarm ran through the darkness. "They move," one hissed. "They choose." "Crush the beacon," another roared. "Smother it before it spreads." But already, the resonance was spreading. On Earth, those who had acted felt warmth wash through them, a wave of affirmation they could not explain. Tears filled the musician's eyes. The scientist's hands trembled. The teacher smiled as the children laughed, certain that someone had answered. On Celestara, Sera Valen's voice broke the silence with a quiet truth. "They are dreaming now. They are choosing." And the convoy pressed

68

onward, their path clearer, their purpose strengthened by the first light of human will.

When one heart is healed, a family is healed; when families heal, nations awaken.

Chapter Fifteen – The Shadows' Retaliation

The Council of Shadows hissed with fury as the fragile beacon refused to break beneath

 their assault. The beacon rose fragile and bright, a trembling thread of light reaching from the Blue World into the void. But even as it formed, the Shadows stirred in reply. Darkness coiled swift and merciless, striking at the light's first breath. On Earth, those who had lit the flame felt the backlash. The teacher who gathered children was mocked by her peers, threatened for planting 'delusions' in young minds. The musician who strummed harmony heard jeers drowning his melody, his song dismissed as madness. The scientist who tuned his instruments found his work sabotaged, his reputation ridiculed. Each faltered, but none let go. Threads of doubt pressed hard, yet a spark of faith kept them steady. In the Council of Shadows, laughter rasped through the dark chamber. "The beacon is fragile," one hissed. "Break it, and the rest will fall." Their voices rose in dissonance, weaving malice into Earth's field, tightening the noose around the light. Far above, Anders' tones rang deep in

Celestara's chamber. "Beacon destabilizing. Shadow interference—direct, sustained. Kael Andersson steadied himself at the helm, vowing silently that the beacon's light would not fail. Andersson's hand pressed firm to the helm. His jaw hardened. "Hold steady. The beacon must not break." Sera Valen closed her eyes, breathing calm into the storm. "Faith is their shield. If they endure, the light cannot be extinguished." The Shadows struck again, lashing at the convoy. Waters and Heart flickered under pressure, their resonance faltering. Kael called the formation harmony. Root anchored crimson. Crown blazed violet. Celestara poured crystalline tones into the breach. Together they shielded the beacon, their unity forming a wall of resonance. For a moment, it seemed not enough. The beacon dimmed, wavering on the edge of collapse. Then, from Earth, new voices rose. Strangers, unknowing of one another prayed in silence, sang in kitchens, meditated in hidden rooms. Across continents, hearts answered, their will adding to the light. The Map flared with sudden brilliance. The beacon surged, brighter than before. The Shadows recoiled, their laughter breaking into snarls of rage. On Earth, the teacher felt warmth in her heart as

71

the children laughed with her instead of at her. The musician's melody carried into the street, where strangers listened in awe. The scientist's instruments pulsed steadily, undisturbed, his hands trembling with wonder. Each knew in that moment—they were not alone. Kael's voice fell into the silence that followed, low and resolute. "The beacon endures. Their choice cannot be unmade." Celestara's thrummed in agreement, her crystalline hull glowing like dawn against the night.

The weight of the message is heavy, but the light that carries it is greater still.

Chapter Sixteen – The Bridge of Light

The convoy shifted into alignment, weaving their harmonics into the first bridge of light.

The world quieted after the storm. Earth's beacon still flickered, fragile yet unbroken, its thread of light holding against the night. In homes and villages, in cities and deserts, dreamers stirred with unease but also with strange anticipation. The Shadows had struck and failed. Something greater was waiting, just beyond the veil. On Earth, more hearts awakened to act. Circles formed in meditation and prayer, their intent rising upward like smoke. Artists painted visions of ships and skies alight with color. Musicians set to song the chords they heard in dreams. Healers laid hands on the weary, their touch humming faint resonance. A scientist stood before his peers, speaking of Voyager's echo, of frequencies no longer random, daring ridicule for the sake of truth. Though scattered, their sparks began to align, weaving threads into a greater resonance. Far above, the convoy felt it. Kael Andersson's gaze lifted from the Map as the convoy began weaving the bridge of light in harmony. "Attune the vessels," he ordered. One by one, the chakra ships aligned—Root glowing red, Waters flowing orange, Solar Plexus blazing gold, Heart radiant green, Throat singing blue, Vision pulsing indigo, Crown shining violet-white. Together they formed a lattice, harmonics interlocking into something greater than the sum of its parts. Anders' tones deepened with awe. "Resonance threshold reached. Bridge forming." The Shadows moved quickly, tendrils writhing, striking at the forming bond. They poured distortion into the rising channel, turning clarity to confusion, hope to doubt. For a moment, the bridge faltered. But the convoy did not yield. Kael steadied Celestara's core. Sera Valen pressed her palms to the Map, her stillness radiating outward. The chakra vessels pulsed brighter, harmonics holding steady, pushing back the tide of distortion. Then, in a blaze of resonance, the bridge pierced the veil. A light brighter than dawn flooded both realms. For the first time, dreamers and convoy crew beheld one another—not as shadows of thought, but in shared vision. On Earth, humans saw the ships with clarity, their colors undeniable. On Celestara, Kael felt the gaze of human souls meeting his own across the gulf of stars. He bowed his head, voice breaking in reverence. "We are no longer strangers." The bridge steadied, glowing with the harmony of choice and will join together. The Shadows recoiled, hissing, their strikes scattering against the resonance now anchored in both Earth and the convoy. At dawn, scattered dreamers awoke with tears of joy. Though worlds apart, each whispered the same words: "The bridge has been lit." On Celestara, Sera's lips curved in a quiet smile. "Now, they will remember." And the fleet pressed onward, the bridge of light a living bond between them and the world they had come to answer.

Time does not only move forward—it circles, returning lessons until they are understood.

Chapter Seventeen – The Shadows' Shroud

The Shadows cast a shroud across the Earth, veiling the bridge in deception and noise.

The bridge glowed steadily across the gulf, a living thread of resonance binding Earth to the convoy, dreamers awoke with clarity, their hearts brimming with visions of ships and light. Artists painted with new fire, musicians played chords they had never known, and circles gathered in prayer, lifting voices into the unseen. The bond was fragile, yet it was real—and both worlds felt it. The Shadows seethed. Their strikes had failed, their coils broken against the convoy's unity. So, they wove an innovative design—not of force, but of concealment. They spun illusions like veils, draping Earth in a shroud of distortion. Lies multiplied, disinformation spread like wildfire, and truths were drowned in endless noise. Dreams were twisted into confusion, leaving souls unsure of what they had seen. On Earth, voices rose in doubt. Some dismissed the visions as fantasy, nothing more than trick of mind. Others mocked the dreamers, calling them fools. The loudest skeptics silenced hopeful voices, and confusion spread wide. Whispers filled the air: "What is real? What is illusion?" On Celestara, Anders' tones carried warning. "Resonance distortion detected. Non-destructive, but obscuring. Visibility reduced." Kael Andersson's eyes narrowed. He understood. "They cannot shatter the bridge. So, they would hide it, make it unseen, until memory fades." Sera Valen's voice flowed soft, steady as the waters. "A bridge hidden is a bridge forgotten. They hope to starve it, not break it. But light need not shout, to be seen. Those who choose to see will pierce the veil." So, the convoy shifted their response. They did not strike back with force. Instead, they amplified subtle harmonics, gentle waves that wove into Earth's field. Small signs appeared: moments of peace where anger had ruled, bursts of kindness in forgotten places, inspiration igniting like sparks. The bridge was not defended by power, but by remembrance. On Earth, a poet wrote of a light in the fog, words flowing as if from beyond himself. A child drew rainbow ships breaking through clouds, her crayon marks glowing with joy. A healer laid hands upon the weary and felt warmth pulsing through her palms. These were threads, fragile but undeniable, piercing the shroud the Shadows had drawn. On Celestara. Kael Andersson narrowed his eyes as the Shadows' shroud wrapped Earth, veiling the bridge in deception.

His words carried quiet resolve. "They cannot unmake what has been chosen." The bridge still shone, veiled but not lost— waiting for every soul who chose to see.

Renewal is not the end of fire, but its true purpose revealed.

Chapter Eighteen – The Call to Unity

Across continents, dreamers began to find one another, their
scattered sparks weaving

toward unity, dreamers began to find one another. A teacher who
had once felt mocked discovered another who shared her visions.
Musicians from different cities played songs with eerily similar
chords. Online circles swelled with accounts of rainbow ships
and harmonics, strangers discovering that their dreams matched.
What once had been isolated whispers became threads weaving
toward one another. Small gatherings grew. Circles of prayer and
meditation widened, hands clasped in silence. Songs were sung
not only to comfort but to remember. Healers joined voices,
lifting their intent together. In one hall, a musician played before
a crowd, and as the chords spread, tears filled eyes unknowing
why. A scientist stood at a podium and dared to speak: "Let us
conduct an experiment in unity. Let us align our thoughts as one,
amplifying the Voyager echo." Far above, Anders' voice carried
into Celestara's chamber. "Increased resonance clustering
detected. Human signals converging Kael Andersson felt resolve

burn within him as unity began to spark across the Blue World. "They are no longer passive. The Blue World is stirring. Its people are becoming participants." Sera Valen's voice, soft yet resonant, followed. "Unity cannot be forced. It must be chosen. And one by one, they are stepping into harmony. This is their true power—the will to join." The Shadows struck back, weaving discord even among the groups. Arguments flared, jealousy whispered, fear stirred. But in several places, compassion prevailed. A quarrel ended with laughter. A jealous word was answered with kindness. Every moment of reconciliation fed the signal, making it stronger than before. The chakra vessels pulsed in rhythm, answering Earth's call. A spiral of light formed, rising and descending, a living resonance stretching between ship and planet. The bridge brightened, woven now not only of dreams but of choice made together. On Earth, across continents, dreamers whispered the same words without knowing why: "We are not alone." On Celestara, Kael bowed his head, his voice deep with certainty. "The call to unity has been answered." And with that call, the convoy pressed onward, no longer bearing light alone, but walking with Earth beside them.

The sound you fear is also the sound that awakens—listen with the heart.

Chapter Nineteen – The Rift of Doubt

Doubt whispered into human minds, threatening to unravel the fragile harmony they had

Built. The bridge shone brighter than ever, a lattice of light threading from Earth into the stars. Circles gathered in prayer, musicians wove songs of vision, dreamers across continents shared accounts of ships and colors. For a moment, unity spread like dawn. The convoy felt it, Celestara thrumming with resonance. But in the shadows, a darker strategy stirred. The Shadows whispered into minds, not with force, but with doubt. Their voices came soft, insidious. "It is only imagination." "You are deluded."

"No one truly stands with you." On Earth, cracks formed. A teacher who had told her students of rainbow ships now wondered if she had planted lies. A musician stared at his guitar, asking if the chords were only invention. A scientist withdrew his call for unity, fearing ridicule more than truth. Groups splintered, arguments rose, and harmony faltered. Far above, Anders' voice carried warning. "Resonance fragmentation detected. Unity field destabilizing. Kael Andersson drew a slow breath as he felt doubt pressing against the bridge, testing every fragile thread.

The bridge flickered, threads unraveling. Force could not restore it. This battle lay in the human heart. Sera Valen's voice flowed steadily, calm as still waters. "Doubt is their sharpest weapon, but faith is stronger. We cannot silence it for them. They must choose."

So the convoy held their harmonics steady, not to intervene, but to keep the space open. They waited. On Earth, a small gathering sat in silence, torn by doubt. Then, from the circle, a child's voice rose clear. "I saw the ships. They are real." The words hung in the air. Some laughed nervously, others frowned. But silence followed, heavy and knowing. In that child's certainty, they felt truth, undeniable. And one by one, their faith returned. The bridge brightened again, threads weaving back into strength. Doubt had struck, but faith reclaimed the song. In the council of Shadows, hisses of frustration echoed. Even their whispers of doubt could not sever what had been chosen. On Celestara,

Kael's voice was quiet, steady. "They stumble, yet they rise. Their faith will hold." The convoy pressed onward, the bridge carrying not perfection, but resilience—the truer strength of unity.

Bridges are built not for escape, but for union between what was and what is to come.

Chapter Twenty – The Harmonic Convergence

The Harmonic Convergence rose as Earth and the convoy sang together for the first time.

Though doubt had tested them, the dreamers rose again, stronger for the trial. Circles regathered, musicians tuned their instruments, mystics opened their hearts in silence, and scientists adjusted their machines with trembling certainty. All felt the same quiet pull, a sense that something was waiting, a moment they must meet up. Across the globe, the call spread. A mother lit a candle at her window, whispering a prayer of welcome. A group of children sang a song of colors they had drawn in crayon. In a hall, voices rose together in harmony, while in hidden rooms, meditators breathed as one. No one commanded them, no one organized them, yet their resonance overlapped, each note flowing into a greater wave. Far above, Anders' tones deepened with awe. "Signal convergence detected. Multiple frequencies align. Synchronization event imminent." The Harmonic Map flared brighter than ever, threads from Earth weaving into a single spiral of light, luminous and undeniable.

Kael Andersson's breath caught as he felt it—not only the convoy guiding Earth, but Earth itself rising to meet them. "They are singing," he whispered. "The planet itself is singing." Sera Valen's eyes shimmered, her voice hushed in reverence. "This is the sound of their becoming." The Shadows surged, hurling discord into the field—anger, fear, and chaos. But the wave of unity drowned them, their dissonance shattered by the overwhelming resonance of choice. The bridge flared unbreakable, woven now from countless human hearts joined in harmony. Then the moment came. A radiant pulse spilled through both realms. On Earth, dreamers across continents saw the fleet clearly, not in fragments or shadows, but in shared vision. On Celestara, the crew felt the press of countless souls, a tide of human presence touching their own. On Earth, a poet bent over his notebook, words flowing as if whispered into his heart: "The song is not theirs alone. It is ours." On Celestara, Kael bowed his head, reverence softening his voice. "The Harmonic Convergence has begun." And the fleet pressed onward, not as guides alone, but as companions in a song now rising together.

Kael Andersson bowed his head in reverence as the Harmonic Convergence rose into full resonance.

Storms may rage, but within them the seed of peace is already planted.

Chapter Twenty-One – The Raising Harmony

Voyager pulsed with unexpected brilliance, no longer a relic but a living beacon of Earth's Voice, amid the rising harmony, a glimmer stirred in the dark. Voyager, the lone wanderer cast out by human hands decades ago, flickered faintly in Celestara's sensors. Once a drifting relic, it now pulsed with light unseen, as though awakened by the call of its creators. Anders leaned forward over the console, his voice taut with awe. "Unexpected resonance spike from Voyager. Signal harmonics aligning with Earth's unity field." Kael Andersson's eyes widened. "It is not just a relic. It is their voice. Their beacon." The golden record within Voyager began to hum with impossible resonance. The greetings, the songs, the sounds of Earth's oceans and winds shimmered into energy. It was no longer mere data but intent— love, curiosity, hope—woven into vibration. A bridge formed, linking humanity's material creation with its spiritual awakening. Sera Valen's words came like a prayer. "They sent it outward as a message of hope. Now, in unity, they have transformed it. Their hope has become light." On Earth, scientists monitoring the deep sky stared at their instruments, baffled. Faint pulses danced near Voyager—patterns too deliberate to ignore. Though they could not explain, awe filled their hearts. Dreamers, too, felt an echo of a star signal answering their prayers. In the shadowed realm, fury rose. The dark ones hissed. "Even their machines betray us." They hurled interference, seeking to choke the beacon, but it held strong. Layered with human will, woven with love, it outshone their discord. Voyager gleamed in the void, no longer alone. Once cast into silence, it now pulsed like a lighthouse, guiding through the deep. On Celestara's bridge, Kael bowed his head, reverence softening his voice. "Earth has lit its beacon. The way is clear." And the convoy aligned its course, guided now not by starlight alone, but by the heart of humanity itself, shining at last into the dark. Kael Andersson's eyes widened as Voyager pulsed with light, no longer a relic but a living beacon.

The hardest gate to pass is not in the heavens, but within your own heart.

Chapter Twenty-Two – The Great Alignment

The chakra fleet locked into perfect formation as the Great Alignment began.

The Harmonic Map, its pulse steady, its voice clear. Across Celestara's bridge, the air itself felt charged waiting, listening. Kael Andersson placed both palms on the helm, guiding the Great Alignment with steady intent. Around him, the chakra vessels held their stations: Root, Waters, Solar, Heart, Throat, Vision, Crown—seven fires awaiting one flame. "Begin the alignment," Kael said. Anders' tones resonated through the chamber. "Initiating harmonic lock. Synchronization across seven channels—phase one." The Root Vessel deepened first, crimson anchoring into the lattice of space like a pillar driven into bedrock. Waters flowed next, orange currents weaving through the pillar, loosening tension, letting movement guide form. Solar kindled gold—will without force, power as service— seeding strength into the weave. Then Heart spoke green: the gentlest tone, yet the strongest, turning the lattice from structure into welcome. Throat rose blue, clearing the channel, giving safe

90

passage to every note. Vision arced indigo, sight cutting a path through the last veils. Crown crowned—violet-white pouring down like dawn, sanctifying the corridor between worlds. On Earth, circles paused as if some unseen conductor had lifted a hand. Candles flickered without wind; breath stilled on countless lips. The moment hung poised choice held like a chord at the edge of resolution. "Phase two," Anders intoned. "Link convoy to beacon."

Celestara answered with a tone so pure it smoothed the bones of the ship. A band of light unfurled from her heart and touched the beacon's pulse. Where light met light, the void brightened. Threads from cities and forests, from lonely rooms and crowded halls, lifted, finding the corridor prepared for them. The bridge widened, no longer a strand but a river. The Shadows struck. Distortion rippled across the map; a black glare meant to blind. It rolled over the corridor—and vanished, absorbed by Waters' flow, grounded by Root's pillar, turned aside by Throat's clarity. What remained broke against Heart like surf on rock.

"Interference dispersing," Anders reported, an edge of wonder beneath his precision. "Their shroud cannot hold within a field of welcome." Kael closed his eyes. He did not command; he listened. He felt the convoy listening, too—each captain laying down the last of their fear, each ship loosening the will to control and choosing, instead, to belong to the song. The alignment moved from numbers to trust. On Earth, a thousand small reconciliations rippled at once: a text sent, a handheld, a silence kept instead of a harsh word. The bridge brightened with every mercy, every honest breath. Children laughed for no reason. Elders wept without shame. A scientist looked up from his monitors and, for the first time, did not doubt what he felt. "Phase three," Anders said softly. "All channels: unite." Seven vessels answered—not louder, but truer. Their hues did not blur; they braided. Where they crossed, new colors appeared—coral and teal, rose-gold and electric violet—shades that were not in any single ship but born of their meeting. The convoy became a single instrument. Celestara became the chamber that held it. Earth became the listening ear—and also, at last, another voice. The chord sounded. It traveled the corridor in a wave that did not push but invite. The beacon brightened until it seemed like a second sun within the Map. Far below, some dreamers saw ships; others saw gardens; some saw only a color they had no word for and felt peace. On the bridge, Kael's throat tightened. He heard within the chord a thousand small lives, and within those lives a decision: not to be saved, but to awaken. The Shadows withdrew, not destroyed but revealed. Their edges frayed where compassion met them; their coils loosened where truth named them. A path opened through their fog—not to flee them, but to walk beyond them. "Alignment complete," Anders said at last. "Convoy locked to Earth's beacon. Corridor stabilized." Kael lowered his hands from the helm. For a long moment he said nothing. Then, quietly: "Hold this. Whatever comes next, hold this." Sera Valen turned toward the map, tears bright. "It will hold," she said. "Because it is held by many." The Great Alignment settled like dawn across the fleet and the world below. Not an ending, but a beginning strong enough to bear all that must follow.

You are never alone; the unseen walks with you in every breath.

Chapter Twenty-Three – Shadows' Retreat

A tidal wave of Shadow interference struck, pressing against the corridor with all its fury.

But the Shadows did not surrender. From the edges of the void they gathered, fury coiling into one final surge. The distortion came like a tidal wave, black and jagged, rolling toward the bridge with force to shatter stars. On Celestara's bridge, Anders' tones sharpened. "Maximum interference surge detected. Amplitude beyond prior limits." Kael Andersson braced as the Shadows' final surge struck like a tidal wave across the corridor. He tightened his grip on the helm, feeling the weight of the strike press against his bones. "Hold the chord. Whatever comes, we do not yield." On Earth, hearts trembled. Fear spread in shadows of thought: *It is hopeless. We are too small. This will break us.* Despair seeped into rooms and dreams. But then, sparks ignited. A mother drew her child into her arms and whispered courage. Strangers lifted one another from rubble after a sudden storm. In temples and kitchens, in forests and alleys, prayers rose. Small acts multiplied, each one a light, and together they flared. The bridge brightened in answer. The chord deepened, no longer fragile but rich, fed not by perfection but by persistence. Sera Valen's voice carried steadily. "The Shadows can only borrow what we give them. Deny them despair, and they starve." The surge struck, but it did not shatter. The field absorbed it, bending and flowing, turning the wave aside like water through stone. The Shadows hissed in rage as their assault faltered. Their darkness broke on compassion's rock, unraveling into tatters. One by one, their coils loosened. Some fled into the outer void; others clung to their edges, weakened, unable to pierce the corridor again. Anders' voice fell quiet, reverent. "Interference collapse. Shadows retreating beyond detection range." Kael's hands dropped from the helm. His voice was soft, but steady. "Then we walk forward, unbound." On Earth, dreamers woke with tears upon their cheeks, hearts lightened by a relief they could not name. A great weight had lifted. The night no longer pressed as heavily as before. The Shadows were not destroyed. But their hold was broken. And the path ahead, for the first time, lay clear.

On Earth, hearts trembled. Fear spread in shadows of thought: *It is hopeless. We are too small. This will break us.* Despair seeped into rooms and dreams. But then, sparks ignited. A mother drew her child into her arms and whispered courage. Strangers lifted one another from rubble after a sudden storm. In temples and kitchens, in forests and alleys, prayers rose. Small acts multiplied, each one a light, and together they flared. The bridge brightened in answer. The chord deepened, no longer fragile

Release is not loss—it is the opening of the soul into freedom.

Chapter Twenty-Four – The Dawn of Gaia

At last, Gaia's dawn shimmered on the Harmonic Map, the path clear and unbroken.

The void lay quiet at last. Where once distortion and shadow churned, the corridor stretched clear—a river of light unbroken between Earth and the convoy. The chakra vessels glowed softly in formation, Celestara at their heart, a constellation suspended in stillness. On Earth, dreamers stirred from sleep with tears they could not explain. Some felt lighter, as though a weight had been lifted from their chests. A child whispered to her parents of gardens she had walked among the stars. A scientist set down his instruments after recording anomalies and, for reasons he could not name, drafted a poem instead. Communities gathered in silence, some around candles, some around fires, some simply beneath the sky. None knew why, only that it felt right. On Celestara's bridge, Kael Andersson stood before the Harmonic Map, heart steady, as Gaia's dawn finally began to rise. He gazed at the Harmonic Map where the beacon pulsed steadily and true. Anders' voice carried calm assurance: "Resonance field stabilized. Beacon continues to guide. No Shadow presence within detection range." Sera Valen's eyes softened, her voice quiet but firm. "This is not the sunrise itself, but the light before it. The first glow upon the horizon." Kael breathed deeply, his gaze fixed on the Blue World below. "They have walked through doubt, through shadow, and yet they stand. We will remain with them. However long it takes, we will not leave until they awaken fully." The chakra ships hummed in agreement, their colors shimmering brighter in the dark. Root and Crown, Heart, and Vision, all aligned, their tones woven now into one. Guardians not of conquest, but of accompaniment. Beneath them, Earth turned slowly, oceans gleaming with sunlight, continents bathed in quiet glow. Clouds parted to reveal green valleys and mountain crowns, as though the world itself leaned into dawn. And in that silence, timeless words seemed to move through both ship and soil alike: The dawn of Gaia had begun. Not an ending, but the first true beginning.

All journeys end where they began—in the embrace of eternal Light.

Epilogue – The Waiting Horizon

The corridor held, steady as dawn light. The convoy drifted in silence, vessels glowing faint against the endless dark. Earth's beacon pulsed steadily, its song carried upward into the stars. The Shadows lingered far beyond detection, their presence dimmed but not erased. For now, the path stands open. On Earth, subtle signs stirred. Awaken souls dreamed of gardens among stars, their sketches uncannily alike. Artists painted ships they had never seen. A farmer in his field felt a joy he could not name. None of them spoke of corridors or convoys, yet each bore a fragment of light awakening within. The dawn spread quietly, like seeds waiting beneath soil. On Celestara's bridge, Kael Andersson stood with Anders and Sera, gazing at the Harmonic Map where Earth's pulse glowed constantly. "The dawn has begun," Kael said softly, "but the day has not yet come." Anders' report came precise yet tinged with reverence. "Beacon stable. Resonance field holding. Shadows remain at distance, observing, waiting." Sera's eyes were calm, luminous with foresight. "The next trials will not come from without, but from within. What has begun in light must grow roots in choice."

Beyond their words, the silence deepened. The Earth turned beneath, oceans shimmering, continents breathing. The fleet sailed onward, not yet finished, not yet triumphant—but faithful to the song rising within the world in the far distance. And a voice beyond time whispered: The journey had not ended. It had only just begun. For Earth stood upon the waiting horizon, between shadow and light, between dream and awakening. The stars held their breath.

Index

Comprehensive Index

www.ingramcontent.com/pod-product-compliance
Lightning Source LLC
Chambersburg PA
CBHW031442120626
46545CB00006B/2516